I0393425

METRIC/ENGLISH CONVERSION FACTORS

ENGLISH TO METRIC

LENGTH (APPROXIMATE)

1 inch (in)	=	2.5 centimeters (cm)
1 foot (ft)	=	30 centimeters (cm)
1 yard (yd)	=	0.9 meter (m)
1 mile (mi)	=	1.6 kilometers (km)

AREA (APPROXIMATE)

1 square inch (sq in, in^2)	=	6.5 square centimeters (cm^2)
1 square foot (sq ft, ft^2)	=	0.09 square meter (m^2)
1 square yard (sq yd, yd^2)	=	0.8 square meter (m^2)
1 square mile (sq mi, mi^2)	=	2.6 square kilometers (km^2)
1 acre = 0.4 hectare (he)	=	4,000 square meters (m^2)

MASS - WEIGHT (APPROXIMATE)

1 ounce (oz)	=	28 grams (gm)
1 pound (lb)	=	0.45 kilogram (kg)
1 short ton = 2,000 pounds (lb)	=	0.9 tonne (t)

VOLUME (APPROXIMATE)

1 teaspoon (tsp)	=	5 milliliters (ml)
1 tablespoon (tbsp)	=	15 milliliters (ml)
1 fluid ounce (fl oz)	=	30 milliliters (ml)
1 cup (c)	=	0.24 liter (l)
1 pint (pt)	=	0.47 liter (l)
1 quart (qt)	=	0.96 liter (l)
1 gallon (gal)	=	3.8 liters (l)
1 cubic foot (cu ft, ft^3)	=	0.03 cubic meter (m^3)
1 cubic yard (cu yd, yd^3)	=	0.76 cubic meter (m^3)

TEMPERATURE (EXACT)

[(x-32)(5/9)] °F = y °C

METRIC TO ENGLISH

LENGTH (APPROXIMATE)

1 millimeter (mm)	=	0.04 inch (in)
1 centimeter (cm)	=	0.4 inch (in)
1 meter (m)	=	3.3 feet (ft)
1 meter (m)	=	1.1 yards (yd)
1 kilometer (km)	=	0.6 mile (mi)

AREA (APPROXIMATE)

1 square centimeter (cm^2)	=	0.16 square inch (sq in, in^2)
1 square meter (m^2)	=	1.2 square yards (sq yd, yd^2)
1 square kilometer (km^2)	=	0.4 square mile (sq mi, mi^2)
10,000 square meters (m^2)	=	1 hectare (ha) = 2.5 acres

MASS - WEIGHT (APPROXIMATE)

1 gram (gm)	=	0.036 ounce (oz)
1 kilogram (kg)	=	2.2 pounds (lb)
1 tonne (t)	=	1,000 kilograms (kg)
	=	1.1 short tons

VOLUME (APPROXIMATE)

1 milliliter (ml)	=	0.03 fluid ounce (fl oz)
1 liter (l)	=	2.1 pints (pt)
1 liter (l)	=	1.06 quarts (qt)
1 liter (l)	=	0.26 gallon (gal)
1 cubic meter (m^3)	=	36 cubic feet (cu ft, ft^3)
1 cubic meter (m^3)	=	1.3 cubic yards (cu yd, yd^3)

TEMPERATURE (EXACT)

[(9/5) y + 32] °C = x °F

QUICK INCH - CENTIMETER LENGTH CONVERSION

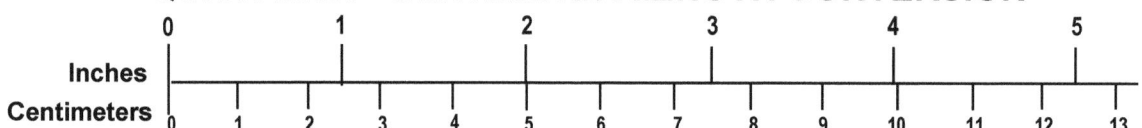

QUICK FAHRENHEIT - CELSIUS TEMPERATURE CONVERSIO

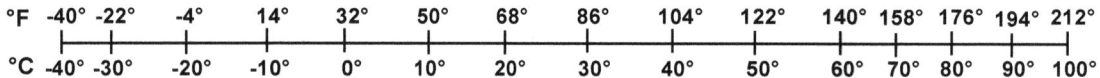

For more exact and or other conversion factors, see NIST Miscellaneous Publication 286, Units of Weights and Measures. Price $2.50 SD Catalog No. C13 10286

Updated 6/17/98

Acknowledgements

The U.S. Department of Transportation (USDOT) Federal Railroad Administration (FRA) Office of Research and Development sponsored the work leading to this report. The authors would like to thank Sam Alibrahim, Chief, FRA Train Control and Communications Division, and Tarek Omar, Program Manager, FRA Office of Research and Development, for their guidance and direction in developing this report.

Special recognition is due to the following organizations for participating in interviews and providing the authors with invaluable information and many of the photographs in this report:

- Amtrak
- Caltrain
- Capital Metro
- San Diego's North County Transportation District
- Metrolink
- New Jersey Transit (NJT)
- Long Island Rail Road (LIRR)

The authors would also like to thank Marco daSilva, Principal Investigator of Surface Transportation Programs, Volpe Center, USDOT John A. Volpe National Transportation Systems (Volpe Center) and Leonard Allen, Chief of the Systems Safety and Engineering Division, Volpe Center, for their leadership and direction.

Contents

Illustrations

vii

Executive Summary

The past two decades have seen an overall reduction in highway-rail grade-crossing accidents and fatalities. Most of that reduction has been with vehicular traffic. Pedestrian accidents and fatalities have more or less held steady.

In 2012, FRA conducted a Right-of-Way Fatality and Trespass Prevention workshop. Pedestrian safety issues and a specific need for research and information dissemination were identified as top research needs by the participants.

This research was conducted to assess innovative and practical solutions for pedestrian grade crossing treatments, to make recommendations and guidelines for future upgrade projects, and to identify additional research for pedestrian treatments for which safety benefit analyses have not been conducted or are not understood.

In addition to researching publicly available information of use of pedestrian grade crossing treatments, an outreach effort was made to each commuter and intercity passenger railroad within the United States. Amtrak, Caltrain, Capital Metro, Coaster, Metrolink, New Jersey Transit, and Long Island Railroad responded and shared some of their solutions to pedestrian safety at railroad grade crossings. Their solutions include inter-track fencing, pedestrian gates, swing gates, channelization, and detectable warnings, as well as other devices (audible warnings, etc.) that are not the focus of this research.

A number of pedestrian treatments have been developed and are used throughout the United States. Many of the treatments are included in the Manual of Uniform Traffic Control Devices (MUTCD), some are being considered for inclusion in the MUTCD, and some are regional or local in use. The pedestrian treatments identified in this document are:

- Smart warning systems
- Pedestrian swing gates
- Detectable warnings / tactile strips
- Directional surfacing
- Pedestrian gates
- Gate skirts
- Flange fillers / surfacing
- Dynamic envelope marking
- Z-crossing (zig-zag)
- Channelization
- Oversized ballast
- Fencing
- Anti-trespass panels

The decision of when to use these treatments is generally a matter of best practices, using the Pedestrian Controls Decision Tree (either the one originally presented in Transit Cooperative Research Program (TCRP) Report 69, or a modified version,) or conducting a site assessment.

9

Research of the efficacy of individual treatments shows that not only do pedestrian automatic gates have the highest level of awareness among survey respondents, but also that they have a stronger effect on actual than stated behavior at reducing pedestrians' risky behavior. Research also shows that while gate skirts were effective in reducing pedestrian gate rushing while the gates were being lowered, their use also resulted in a 12 percent increase in gate rushing when the gates were ascending. Overall though, research on the efficacy of pedestrian treatments at highway-rail grade crossings is limited. Most research has studied the effect of a particular individual pedestrian treatment. Pedestrian treatments, however, are seldom if ever used in isolation. Combinations of pedestrian treatments should also be studied for their efficacy in reducing pedestrian risky behavior.

In 2010 the Secretary of Transportation called on transportation agencies to improve opportunities for pedestrian and bicyclists, foster their increased use of transportation facilities, and to go beyond the minimum requirements. The Proposed Guidelines for Pedestrian Facilities in the Public Right-of-Way impact future design considerations for pedestrian railroad grade crossings. In anticipation of more non-motorized users of varying abilities making use of pedestrian grade-crossing facilities, it is more important than ever that the efficacy of pedestrian treatments at grade crossings be fully understood.

1. Introduction

The John A. Volpe National Transportation Systems Center (Volpe Center) provides technical support to the U.S. Department of Transportation (USDOT) Federal Railroad Administration (FRA) Office of Research and Development on issues involving railroad safety and trespass prevention. The aim of this effort is to produce a body of information that will assist current and future researchers in improving rail safety.

Over the past two decades, federal, state, and local governments, and railroads, have worked towards and succeeded in reducing accidents and fatalities at highway-rail grade crossings. However, much of that benefit is experienced by automotive traffic, with pedestrian incidents and fatalities holding steady.

Figures 1 and 2 illustrate the twenty year trend of incidents and fatalities, respectively, at highway-rail grade crossings in the United States.[1] From 1994 through 2013 there has been a 57% decrease of vehicular incidents at highway-rail grade crossings, along with a 69% decrease in fatalities. Pedestrians have not benefitted from the same decrease; there were 50 pedestrian fatalities at highway-rail grade crossings in 1994 and 65 in 2013 [1].

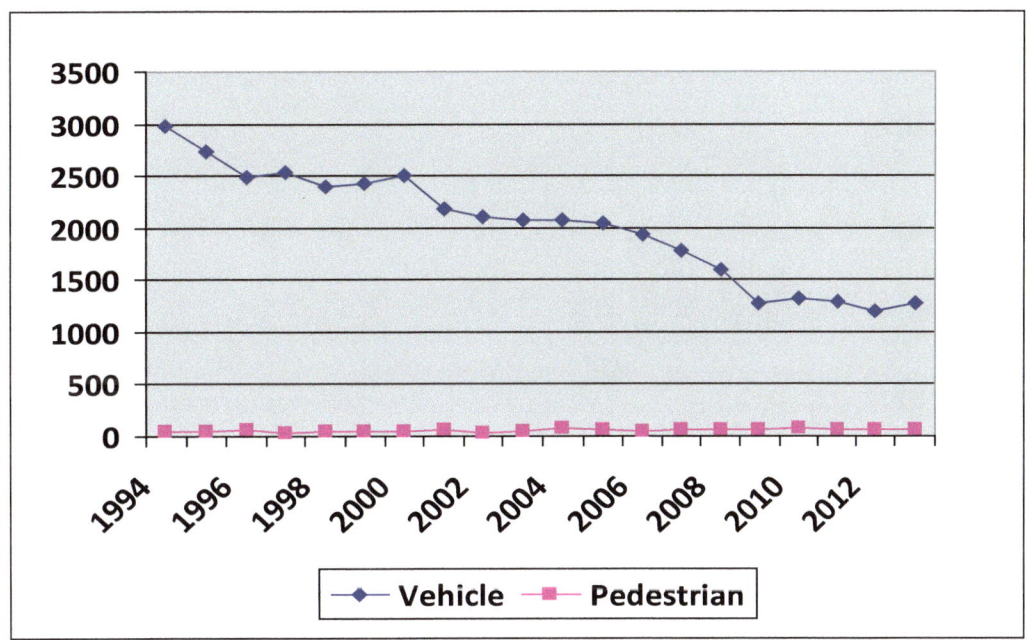

Figure 1. Vehicle and Pedestrian Incidents at Grade Crossings, 1994 – 2013 [1]

---

[1] Data was taken from FRA Safety data (http://safetydata.fra.dot.gov/OfficeofSafety/default.aspx). Vehicular data includes automobiles, pick-up trucks, vans, buses, and school buses.

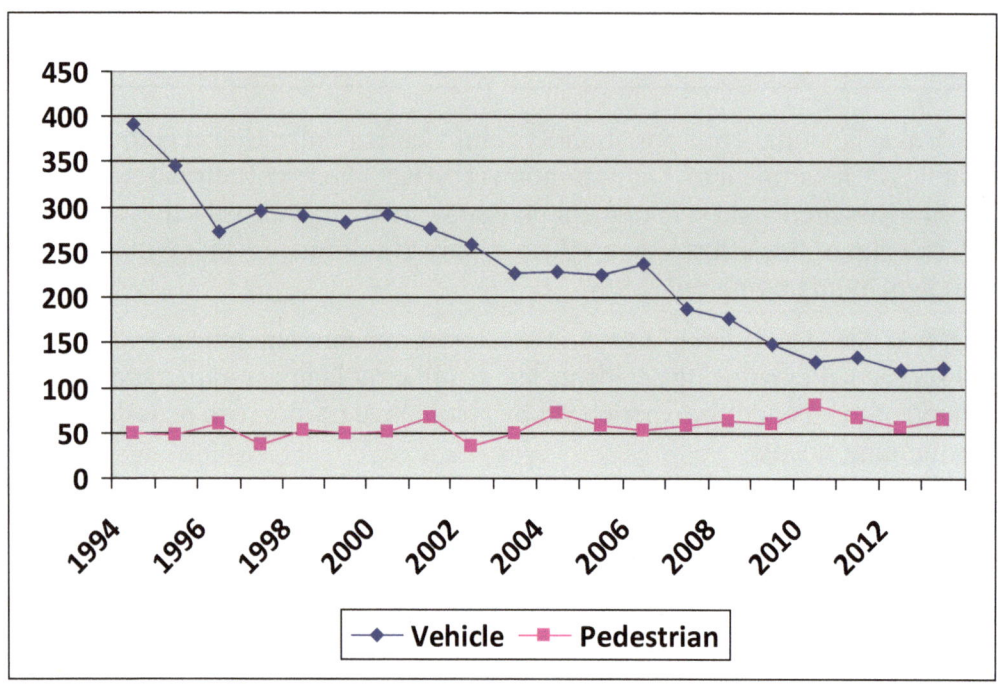

Figure 2. Vehicle and Pedestrian Fatalities at Grade Crossings, 1994 – 2013 [1]

1.1 Background

Risky pedestrian behavior at actively protected highway-rail grade crossings continues to be a significant problem for the railroad industry. Overall pedestrian safety issues and a specific need for research and information dissemination were deemed top research needs by the participants at the 2012 Right-of-Way Fatality and Trespass Prevention Workshop [2].

1.2 Objectives

The purposes of the research reported here are to assess innovative and practical solutions for pedestrian grade crossing treatments, to make recommendations and guidelines for future upgrade projects, and to identify additional research for pedestrian treatments for which safety benefit analyses have not been conducted or are not understood.

1.3 Overall Approach

To document information of pedestrian grade crossing treatments currently in use, a number of investigation techniques were used, including a web search, a literature review, and industry outreach to rail safety stakeholders.

First, a web search was performed to visually inspect and collect images of the types of pedestrian treatments currently in use. Both domestic and international treatments were considered for inclusion in this report. The search was made using terms such as: pedestrian grade crossing, grade crossing, pedestrian treatments, pedestrian rail-grade crossing safety, etc.

A literature review was also performed to discover the types of pedestrian safety treatments used by railroads, as well as the decision-making tools used by engineers in the design of pedestrian rail grade crossings, any studies regarding the efficacy of treatments currently in use, as well as any other pertinent information regarding the use of pedestrian safety treatments at rail-grade crossings.

Finally, an outreach effort was made to each commuter and intercity passenger railroad within the United States. A letter describing the project, and asking railroads to offer the solutions they use to ensure pedestrian safety and rail grade crossings, was sent to 26 passenger railroads. Amtrak, Caltrain, Capital Metro, Coaster, Metrolink, New Jersey Transit, and Long Island Railroad responded and shared some of their solutions to pedestrian safety at railroad grade crossings. Their solutions included inter-track fencing, pedestrian gates, swing gates, channelization, and detectable warnings. These agencies use other devices as well that are not the focus of this research, such as audible warnings and signage.

1.4 Scope

The scope of this study is to research and compile various treatments and strategies used to provide for pedestrian safety at rail-grade crossings, with a particular eye for cost-effective or innovative strategies and treatments. The end product is a compilation of various treatments and recommendations for future research.

The scope of this investigation is limited to the review of publicly available technical documents, results of discussions with specific transportation agencies, and review of localities identified within the body of this report. These agencies were identified by informal outreach via a "Request for Information" (RFI), as shown in Appendix A. This sampling is by no means an exhaustive list, and only represents results from available documents, agency interviews, and locations that were visited.

1.5 Organization of the Report

This report is organized into the following chapters:

- The first is this introduction.

- The second presents the findings from the literature review and outreach efforts.

- The third presents the pedestrian treatments currently in use.

- The fourth discusses the risky pedestrian behavior, efficacy of pedestrian grade-crossing treatments and the decision process that goes into specific treatments' use.

- The fifth discusses other considerations

- The sixth and final chapter summarizes the conclusions and recommendations.

2. Literature Review and Outreach

2.1 Literature Review

Three documents set the recent standards for current best practices for pedestrian treatments at rail grade crossings, the USDOT Federal Highway Administration (FHWA) Manual on Uniform Traffic Control Devices (MUTCD) last published in 2009 [3], the FHWA Railroad Highway Grade Crossing Handbook, published in 2007 [4], and the USDOT FRA's Compilation of Pedestrian Safety Devices in Use at Grade Crossings, published in 2008 [5].

Part 8 of the MUTCD outlines direction for Traffic Control for Railroad and Light Rail Transit Grade Crossings. The handbook recommends passive and active devices as potential treatments for pedestrian rail grade crossings. Passive devices include fencing, swing gates, pedestrian barriers, pavement markings and texturing, etc. Active devices include automated pedestrian gates, variable message signs, etc.

The Compilation of Pedestrian Safety Devices in Use at Grade Crossings illustrates a number of treatments, both compliant and non-compliant with the FHWA MUTCD. Among those are safety devices used to focus attention, such as pedestrian swing gates and "zigzag" or "Z-Gate" crossings. The document claims that swing gates have contributed to reduced incidents related to passenger inattention.

Mathieu identifies early grade crossing treatments as "sidewalk enhancements" and illustrates two in his presentation [6]. Each illustration shows pedestrian treatments virtually ending at the grade crossing. The presentation goes on to identify the Metrolink Astoria Street crossing zig-zag fencing as a 1990's pilot project that encouraged pedestrians to look both ways. There are no recorded incidents at the crossing as of the date of his 2007 presentation.

A California Public Utilities Commission (CPUC) report, Pedestrian-Rail Crossings in California [7], compiled a list of design elements commonly used at pedestrian rail-grade crossings in California. These treatments are:

- Swing Gates
- Detectable Warnings (tactile strips)
- Pedestrian Gates
- Flashing Light Signal Assembly
- Signage
- Crossing Surfacing
- Channelization Design

The Transportation Research Board (TRB) Transit Cooperative Research Program (TCRP) Report 69 presented a decision tree that defined the type of pedestrian devices and controls that are desirable based on six criteria (decision points) relative to the pedestrian crossing environment [8].

The decision tree criteria are as follows:

14

- Pedestrian facilities and minimum pedestrian activity present or anticipated
- Light Rail Transit (LRT) speed exceeds 35 miles per hour
- Sight distance restricted on approach
- Crossing located in school zone
- High pedestrian activity levels occur
- Pedestrian surges or high pedestrian inattention

Caltrain recommends, in addition to pedestrian gate arms with emergency swing gates, pavement texturing and marking, channelization (including guard railing and fencing), and swing gates [9]. Furthermore, a site specific evaluation should be performed if any of the following conditions exist:

- Adjacent to a station
- Adjacent to or near a school or senior center
- Adjacent to or near dense residential or commercial attractions
- High volume pedestrian traffic

Pedestrian grade crossing safety stakeholders are awaiting the release of TCRP Project A-38, being conducted by the Texas A&M Transportation Institute (TTI) [10]. This research is expected to address safe and effective pedestrian crossing treatments and operating practices for rail public transit that can be used with greater consistency across the country.

2.2 Outreach

Each commuter rail and intercity passenger railroad, or the applicable passenger railroad contractor responsible for operation and maintenance of the railroad, was contacted in the Request for Information outreach. An example of the letter sent is shown in Appendix A. The RFI informed the railroad of the project intent and also outlined the type of information sought. From each railroad the following was requested:

- The types of pedestrian safety engineering and design criteria used,
- The types of engineered pedestrian treatments currently in use, and
- Any data to the effectiveness of those treatments.

Amtrak

Amtrak provided a detailed summary of pedestrian safety features along the Northeast Corridor, New England Division in Connecticut and Rhode Island. Amtrak is very proactive in ensuring that inter-track fencing has been installed and maintained at all passenger stations. The inter-track fencing is installed at least the length of the station platform to impede passengers from crossing over the tracks where it is not permitted or safe to do so. In areas where a highway-rail grade crossing is next to a station, pedestrians are routed to the crossing to safely cross the tracks to the opposite side. In areas that do not have a highway-rail grade crossing, overhead or underground (tunnel) passageways that are American with Disabilities Act (ADA) compliant are provided.

Another important safety feature on the Northeast Corridor, New England Division is that of the Train Approaching Message Systems (TAMS). This visual and audible warning device alerts passengers when a train is approaching the station, either to stop or to run through the station. It activates a minimum of 20 seconds before the arrival of a train and stays activated until the train has left the station.

Amtrak also promotes pedestrian safety by using education. Operation Lifesaver (OL) has been active in promoting pedestrian safety in all States including those in New England. For example, ten years ago the Rhode Island OL chapter received ten used computers and placed them in schools and community centers. These computers were used to deliver a kiosk-type power point presentation on the dangers of trespassing. The computers were able to operate in a stand-alone mode, without the need to have an authorized presenter on site. This enabled OL to get the message out whenever the facility was open. The program was a success for Rhode Island. Over time, however, the computers broke and became obsolete, there was no funding to replace them, so the program ceased.

The Connecticut OL chapter, along with the Massachusetts chapter, is equally active in promoting pedestrian safety. Several Amtrak Police officers in Connecticut are OL presenters. These officers maintain professional relationships with local judicial officials, and have established a dialog with them such that, when a person is charged with trespassing on railroad property, the offender must attend an OL presentation with one of the Amtrak police OL presenters.

Trespassing on Amtrak property is mainly addressed by public outreach and enforcement initiatives using state, local and railroad resources. Intensity and sustainability of such campaigns varies by locale as it usually requires a strong local advocate to maintain such efforts in the long run [11].

Caltrain

Caltrain provided its grade crossing design criteria contained in standard drawings for several types of grade crossings [12]. Caltrain uses pedestrian gate systems, swing gates, channelization and signage to control pedestrian safety at grade crossings. Figure 3 through Figure 6 present some examples provided by Caltrain [13].

Figure 3. Example of Pedestrian Gate with Emergency Exit Gate [13]

Figure 4. Example of Inter-track Fencing at a Station [13]

Figure 5. Example of Pedestrian Overpass at a Station [13]

Figure 6. Example of Inter-track Fencing at a Low Level Platform Station [13]

Capital Metro

Capital Metro provided its grade crossing design criteria. Capital Metro uses pedestrian gate systems, swing gates and channelization and signage to control pedestrian safety at grade crossings [14].

Coaster

San Diego's North County Transit District (NCTD) provided a detailed PowerPoint presentation titled "NCTD Grade Crossing Pedestrian Treatments" [15] that covered its SPRINTER Light Rail operation and COASTER Commuter Rail Operation. Most of the Sprinter line grade crossings feature yellow detectable warnings, automated bells and lights and "Look Both Ways" signage. On the Coaster commuter rail line, the Carlsbad Station features automated lights and bells with no arms, high speed train danger signs along the west wall next to the tracks, yellow lines with nonskid surfaces, two yellow swing gates, pedestrian prohibited signage along the east side of the track, and yellow detectable warning on all quadrants of the Grade Crossing. Figure 7 through Figure 10 show some examples from the referenced presentation [16].

Figure 7. Example of Yellow Detectable Warning Strips [15]

Figure 8. Example of a Z-Crossing at a Railroad Crossing [15]

Figure 9. Example of a Pedestrian Crossing with Swing Gates [15]

Figure 10. Example of Bollards and Fencing as Channelization [15]

Metrolink

Metrolink provided the research team with a detailed Highway-Rail Grade Crossings – Recommended Design Practices and Standards Manual. Metrolink uses pedestrian gate systems, swing gates, channelization and signage to control pedestrian safety at grade crossings [17]. Examples of two treatments are shown below.

Figure 11. Pedestrian Gates and Swing Gates in Use at a Grade Crossing [17]

Figure 12. Example Pedestrian Second Train Warning Signage [17]

New Jersey Transit

New Jersey Transit provided two FRA/Volpe Center research reports that describe some best practices with fencing, channelization, signage and second train warning devices [18]. Those reports are:

> *Effect of an Active another Train Coming Warning System on Pedestrian Behavior at a Highway-Rail Grade Crossing* [19]

> *Effect of Gate Skirts on Pedestrian Behavior at Highway-Rail Grade Crossings* [20]

Some of the treatments in use by New Jersey Transit are shown in Figures 13 through 15.

Figure 13. Example of Pedestrian Gate with Gate Skirt [20]

Figure 14. Example of Channelization at a NJ Transit Railroad Station [18]

Figure 15. Example of Warning Signage on NJ Transit Inter-track Fencing [18]

Long Island Railroad

The Long Island Railroad (LIRR) provided pictures showing three pedestrian safety treatments at grade crossings: the Light Emitting Diode (LED) lighting/four quadrant gate system, electronic bell replacement for grade crossings, and loop sensor technology embedded in the pavement of grade crossings [21]. Photos of each of these treatments are shown in Figure 16 through 18.

Figure 16. A LED Lighted Quadrant Gate System [21]

Figure 17. Electric Bell at a Grade Crossing [21]

Figure 18. Loop Sensor Technology Being Embedded at a Grade Crossing [21]

3. Pedestrian Treatments

The goal of this section is to gather all of the pedestrian grade crossing treatments being used throughout the United States. Personal interviews, industry outreach, literature reviews, and web searches were all used to gather the information.

Each section provides a technical description of the treatment, as well as a picture or illustration. Additional information includes

- Whether the treatment's use is optional or mandatory, as outlined in the MUTCD.

- The treatment's MUTCD status. Many of treatments presented are contained within the MUTCD, version 2009. Some treatments are in the process of being vetted for inclusion in the MUTCD. Others may be novel approaches or unique situations whose treatments may never be presented for inclusion into the MUTCD.

- Cost - estimated or actual costs of warning systems were generally not readily available [22]. The authors used their best professional judgment regarding the relative cost of a treatment, and whether that cost was low, medium, or high.

- Places where the pedestrian treatment is known to be in use.

The treatments identified during this study are:

- Smart warning systems
- Pedestrian swing gates
- Detectable warnings and tactile strips
- Directional surface
- Pedestrian gates
- Gate skirts
- Flange fillers and surfacing
- Dynamic envelope markings
- Z-crossing (zig zag)
- Channelization
- Oversized ballast
- Bedstead barriers
- Fencing
- Anti-trespass panels

3.1 Smart Warning Systems

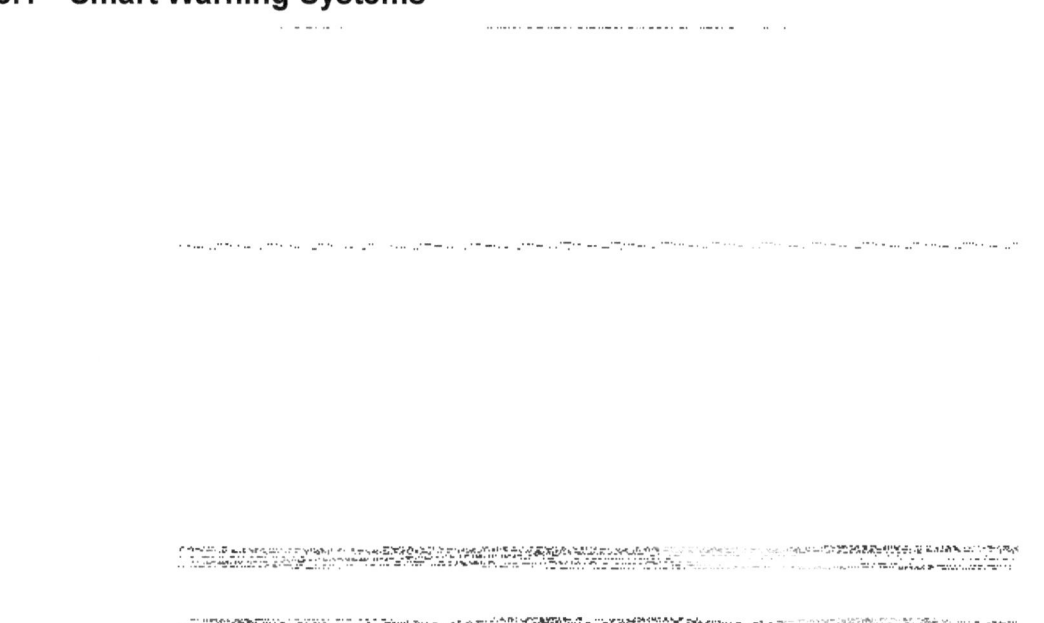

Figure 19. An SWS Sign Displays "Train Approaching" Message [11]

Description:

Smart Warning Systems (SWS) utilize video monitoring of a railroad crossing or station platform, available 24 hours per day with special low light capability and illumination. The system uses advanced multispectral imaging, image-based metrology, and unique algorithmic approaches to determine the size and nature of any object intruding on the defined "exclusion region" around the tracks. Objects intruding on the region are tracked and given a cautionary yellow marker; red markers are objects in the exclusion zone which have stopped in the region and remain there. Smart processing software will automatically alert any interested parties – including central dispatch, local emergency authorities, and even the train if properly equipped – if something remains on the tracks too long or fits a potentially dangerous profile when a train is approaching. This may be tied into notification systems to warn pedestrians of approaching movements.

MUTCD Status: Not applicable

Optional or Mandatory: Optional design feature.

Cost: Low/moderate cost to install & maintain

Place(s) Where Treatment Known to be Used: Used exclusively on the Northeast Corridor from Boston to Washington D.C.

3.2 Pedestrian Swing Gates

Figure 20. Example of Pedestrian Swing Gates at a Grade [7]

Description:

The swing gate (sometimes used in conjunction with flashing lights and bells) alerts pedestrians to the tracks that are to be crossed and forces them to pause, thus deterring them from running freely across the tracks without unduly restricting their exit from the right of way. The swing gate requires pedestrians to pull the gate to enter the crossing and push the gate to exit the protected track area; therefore, a pedestrian cannot physically cross the track area without pulling and opening the gate. Transit Cooperative Research Program Report 17 (TCRP Report 17) recommends that the gates be designed to return to the closed position after the pedestrian has passed [23].

MUTCD Status: Not Applicable

Optional or Mandatory: Optional design feature

Cost: Low/Moderate Cost to Install & Maintain

Place(s) Where Treatment Known to be Used: Caltrain, Metrolink, NCTD Coaster, Altamont Commuter Express, Los Angeles County Metropolitan Transportation Authority (LACMTA), Amtrak Capital Corridor

3.3 Detectable Warnings and Tactile Strips

Figure 21. Example of Detectable Warnings at a Transition Point in the Pedestrian Pathway [7]

Description:

Detectable warnings are a distinctive surface pattern of domes detectable by cane or underfoot that alert people with vision impairments of their approach to street crossings and hazardous drop-offs. They are used to indicate the boundary between pedestrian and vehicular routes where there is a flush instead of a curbed connection. Detectable warnings also indicate unprotected drop-offs along the edges of boarding platforms at stations.

Considerations:

The Access Board is currently finalizing the guidelines for public rights-of-way and shared use paths based on the public comments received on the proposed version. The proposed rule offers an interim source of guidance on access to public rights-of-way, including the provision of detectable warnings, until the final guidelines are published. Once finalized, the guidelines, though usable, will not be mandatory until implemented as enforceable standards by other agencies, including DOT.

MUTCD Status: Included in standard

Optional or Mandatory: Mandatory once enforceable standards are issued [24].

Cost: Low Cost to Install & Maintain

Place(s) Where Treatment Known to be Used: Nationwide

3.4 Directional Surface

Figure 22. Directional Surface Tiles Embedded in Walking Surface Lead Visually Impaired Pedestrians to Safe Grade Crossing Locations [7]

Description:

Similar to Detectable Warnings and Tactile strips, Directional Surface Tiles indicate way finding information to pedestrians with vision impairments. The channelization strips provide orientation clues to the designated pathway over the tracks. They can also be used to direct pedestrians with vision impairments away from unprotected drop-offs along the edges of boarding platforms at stations.

MUTCD Status: Included in standard

Optional or Mandatory: Will be mandatory once enforceable standards are issued [24].

Cost: Low Cost to Install & Maintain

Place(s) Where Treatment Known to be Used: Nationwide

3.5 Pedestrian Gates

Figure 23. Pedestrian Gates in the Upright Position at a Grade Crossing [5]

Description:

Pedestrian automatic gates are the same as standard automatic crossing gates except that the gate arms are shorter. When they are activated by an approaching train, the automatic gates are used to physically prevent pedestrians from crossing the tracks.

MUTCD Status: Included in standard

Optional or Mandatory: Installed in most cases at grade crossings with pedestrian sidewalks

Cost: Moderate

Place(s) Where Treatment Known to be Used: In use nationwide on passenger rail corridors as well as freight and light rail corridors

3.6 Gate Skirts

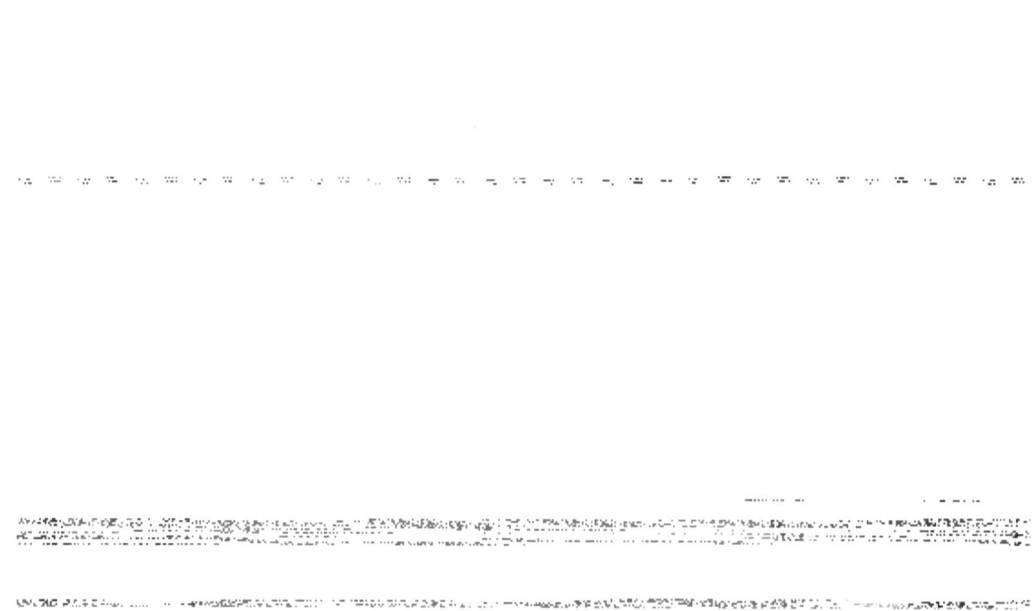

Figure 24. A Gate Skirt in the Deployed Position [20]

Description:

A gate skirt is a secondary bar (or gate) marked with identical striping to the primary gate arm. The skirt hangs down from the pedestrian gate, and, in its fully deployed horizontal position, blocks additional area under the gate arm.

Gate skirts are thought to be a beneficial addition to crossings, especially where there is evidence of pedestrians going under existing gates, or at crossings that many children use. Gates skirts theoretically make it more difficult for pedestrians to violate the crossing after the gates have fully descended, especially at crossings where some barrier channelization exists.

MUTCD Status: Under consideration as an additional safety treatment.

Optional or Mandatory: Installed in most cases at grade crossings with pedestrian sidewalks.

Cost: Low

Place(s) Where Treatment Known to be Used: New Jersey Transit, Dallas Area Rapid Transit

3.7 Flange Fillers and Surfacing

Figure 25. Bicyclists Easily Traverse Tracks Where Flange Fillers are in Use [25]

Description:

The flange way gap is the open area adjacent to the head of the rail to permit the wheel flanges to pass. A flange filler is a rubber insert that fills the flange way gap providing pedestrians and other crossing users with a level, gap free surface. The insert will deflect downwards with the weight of a train.

Currently there are no design strategies that can completely eliminate the flange way gap for high speed passenger and freight rail systems. The gap could be eliminated in the future with further research to develop a system similar to what is currently available for low speed, light rail trains.

MUTCD Status: There are no MUTCD compliant flange fillers.

Cost: Low

Place(s) Where Treatment Known to be Used: Light Rail or Commuter Rail/Intercity rail Rights-of-way (ROWs) that utilize street running.

3.8   Dynamic Envelope Marking

Figure 26. Dynamic Envelope Marking in an Application Geared Towards Automotive Traffic [26]

Description:

The dynamic envelope is defined as the clearance required for the train and its cargo overhang due to any combination of loading, lateral motion, or suspension failure. Dynamic envelope markings are highly visible pavement markings along the right-of-way, usually following the "Do Not Block Intersection Markings" section of the MUTCD. A Dynamic Envelope Marking study for vehicles was conducted in Ft. Lauderdale in 2012-13 using the markings design shown in Figure 26. The results of the study were encouraging, with a 45% reduction in vehicles stopped on the tracks and an overall reduction in gate violations. There have not been any studies on the effectiveness of using dynamic envelope markings for pedestrian treatments.

MUTCD Status: Included in standard

Cost: Low/Moderate

Place(s) Where Treatment Known to be Used: Ft. Lauderdale, Florida (design shown above), and nationwide for other design patterns mostly consisting of white cross-hatching lines.

3.9 Z Crossing (Zig-Zag Crossing)

Figure 27. Example of Z-crossing in the City of Lemon Grove, California [7]

Description:

Z-crossings are designed to channel pedestrians in such a manner that they are forced to look down the tracks while approaching the crossing

MUTCD Status: Included in standard

Cost: Low/Moderate Cost

Place(s) Where Treatment Known to be Used: California, New Jersey

3.10 Channelization

Figure 28. Swing Gates in Combination with Fencing Used to Achieve Channelization [27]

Description:

Channelization is a technique that provides clear, well-defined pathways for pedestrians to cross the tracks where warning devices are in place. Channelization also discourages improper behavior, such as circumventing the gates or walking onto the right-of-way. Channelization may include fencing, guard railing, swing gates, and various control devices.

Physical channelization using fencing is critical to the effectiveness of pedestrian gates and swing gates because it prevents pedestrians from easily circumventing the devices. A study performed in Illinois [9] demonstrated that pedestrians regularly violated pedestrian gates at crossings that did not include adequate channelization as a design element.

Pathway delineation and directional signage may assist in channelization, particularly at places where physical fencing cannot be provided such as at the edge of a station platform or at the track surface. Delineation of the pathway can be provided by white edge line markings or contrasting pavement color or texture.

Figure 29. Example of a Z-crossing Using Fencing to Achieve Channelization [28]

Considerations: Use is optional and site specific.

MUTCD Status: Included in standard.

Cost: Low/Moderate

Place(s) Where Treatment Known to be Used: California, New Jersey

3.11 Oversized Ballast

Figure 30. Oversized Ballast Along a Rail Corridor Encourages Pedestrians to Cross at Approved Pedestrian Crossing [29]

Figure 31. Oversized Ballast Along a Ramp Discourages Pedestrians from Taking a Shortcut [7]

Description:

Oversized ballast is used to prevent unauthorized access in the area near the ROW and stations.

MUTCD Status: Not applicable.

Optional or Mandatory: The use of oversized ballast is optional.

Cost: Low Cost

Place(s) Where Treatment Known to be Used: Nationwide

3.12 Bedstead Barriers

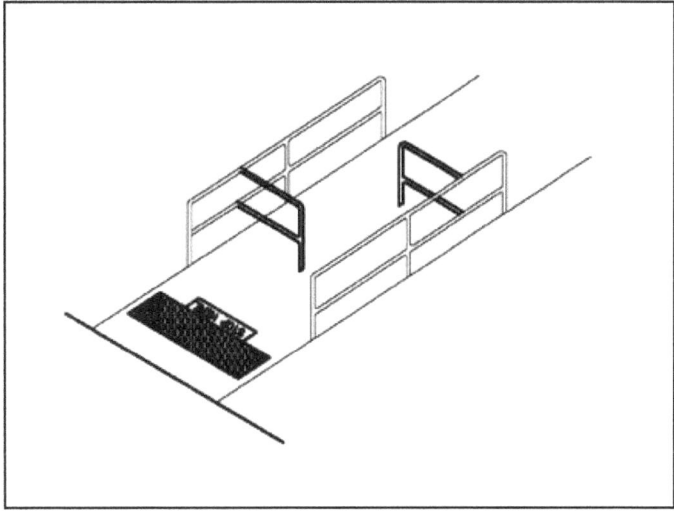

Figure 32. Illustration of a Bedstead Barrier [30]

Description:

The bedstead barrier concept may be used in tight urban spaces where there is no fenced-in right of way. The barricades are placed in an offset (maze-like) manner that requires pedestrians moving across the tracks to navigate the passageway through the barriers. TCRP Report 17 [23] recommends that they be designed and installed to turn pedestrians towards a rail vehicle's approach before they cross each track, forcing them to look in the direction of an oncoming vehicle. Report 17 also recommends bedstead barriers for crossings where pedestrians are likely to run unimpeded across the tracks where the risk of collision is low to medium, and may be used with automatic gates in high-risk areas. The Report also recommends bedstead barriers not be used where vehicles operate in both directions on a single track because pedestrians may look in the wrong direction in some instances. Bollards and chains may be used to accomplish the same effect as bedstead barriers.

MUTCD Status: Not applicable

Optional or Mandatory: Optional

Cost: Low

Place(s) Where Treatment Known to be Used: Utah

3.13 Fencing

Figure 33. Standard Fencing Along a Right of Way [13]

Figure 34. Example of Newer Style of Fencing with Shorter Spacing to Deter Climbing and Fence Cutting [13]

Description:

Different types of fencing materials, such as wire mesh and steel tubular fencing, are used to separate the railroad ROW from a highway or other property lines.

MUTCD Status: Included in standard

Optional or Mandatory: Mandatory

Cost: Low (Maintenance can be high)

Places in Use: Nationwide

3.14 Anti-Trespass Panels

Figure 35. Projecting Cones Adjacent to Pedestrian Grade Crossing [31]

Description:

Anti-trespass panels are made of rubber and feature projecting cones. Placed adjacent to pedestrian grade-crossings, these panels may provide visual and tactile deterrence to keep pedestrians on the intended path through a grade-crossing.

MUTCD Status: Not included in standard

Optional or Mandatory: Optional

Cost: Low

Places in Use: Europe, Japan

4. Pedestrians and Grade Crossing Treatment Design

4.1 Grade Crossing Treatments and Pedestrian Behavior

Pedestrians have historically demonstrated risky behavior at grade crossings. However, new treatments to mitigate risky behavior often result in new types of risky behavior. The most basic version of this is when pedestrian gates were installed at pedestrian crossings because of pedestrians disregard for flashing lights and audible warnings. When gates were lowered, some pedestrians simply went around them rather than waiting for the train to pass.

Between 2003 and 2007 non-motorist incidents at grade crossings remained unchanged while highway-rail grade crossings declined 44 percent between 1994 and 2007. It is thought that this is because safety measures, such as gates and flashing lights, focus mainly on motorists [32]. It is not surprising, then, that since 2007 there has been a significant increase in the installation of engineered pedestrian treatments.

Actual, as opposed to stated, pedestrian behavior at grade-crossings has been the subject of several studies. Metaxatos revealed the tendency of pedestrians and bicyclists to pass around fully lowered gates was greater when they were crossing the tracks in groups rather than when they were alone [33]. This phenomenon is also known as "platooning" [34]. An example of this type of behavior is shown in Figure 36 below.

Figure 36. Multiple Pedestrian Violations at an Active Grade Crossing in Little Neck, NY [34]

Siques showed that the use of pedestrian treatments can result in risky behavior among pedestrians entering a railroad crossing [32]. Specifically, automatic gates at Light Rail Vehicle (LRV) crossings resulted in statistically significant improvements in risky pedestrian behavior and pedestrian channelization reduced the number of pedestrians deviating from the sidewalk. However, in the same research, pedestrians also revealed that risky behavior (less likely to look both ways or to stop) can increase in the presence of pedestrian automatic gates or active

42

warning devices, presumably because the warning devices take the decision making away from them [31] .

Metaxatos' research showed presence of automatic gates had elicited the highest level awareness from survey respondents. His research indicated that automatic pedestrian gates had an even stronger effect of deterring actual as opposed to stated behavior [22].

A pilot study measuring the effectiveness of gate skirts on pedestrian behavior showed that, while the skirts were effective at reducing gate-rushing by pedestrians when the gates were lowering, gate-rushing actually increased by 12 percent when the gates were rising [20].

These studies provide some evidence of pedestrian behavior with regard to engineered treatments at grade-crossings, but there is limited empirical evidence on what treatments, or combinations of treatments, work the best to reduce pedestrian risky behavior and assure their safety and rail-grade crossings.

4.2 Determining Pedestrian Treatments

The decision of which treatments to use at a pedestrian grade-crossing is currently based on a combination of site assessments, best practices, and decision trees. A number of approaches to determining the best treatments for a grade crossing have been developed.

The TCRP developed the decision tree shown in Figure 37 as a recommended practice in TCRP Report 69 [8]. That decision tree is currently in use by CPUC and is incorporated into its compilation of pedestrian rail-grade crossing designs and devices document [7].

Capital Metro Railroad (CMTA) also developed a consideration flowchart for engineers to use during the design of pedestrian-rail grade crossings. The flowchart takes proximity to hospitals and "ADA facilities" into consideration, as well as further analysis of multiple track situations, such as whether or not there are siding or main tracks. Based on the features of the grade crossings, the recommendation would be either standard passive, full standard, or full standard plus either access channelization, swing gates, active warning devices, or grade separation [35].

In addition, the design has been adapted and adopted by a number of other rail operators, such as Metrolink [27], and in is in use in Denver, Colorado [36].

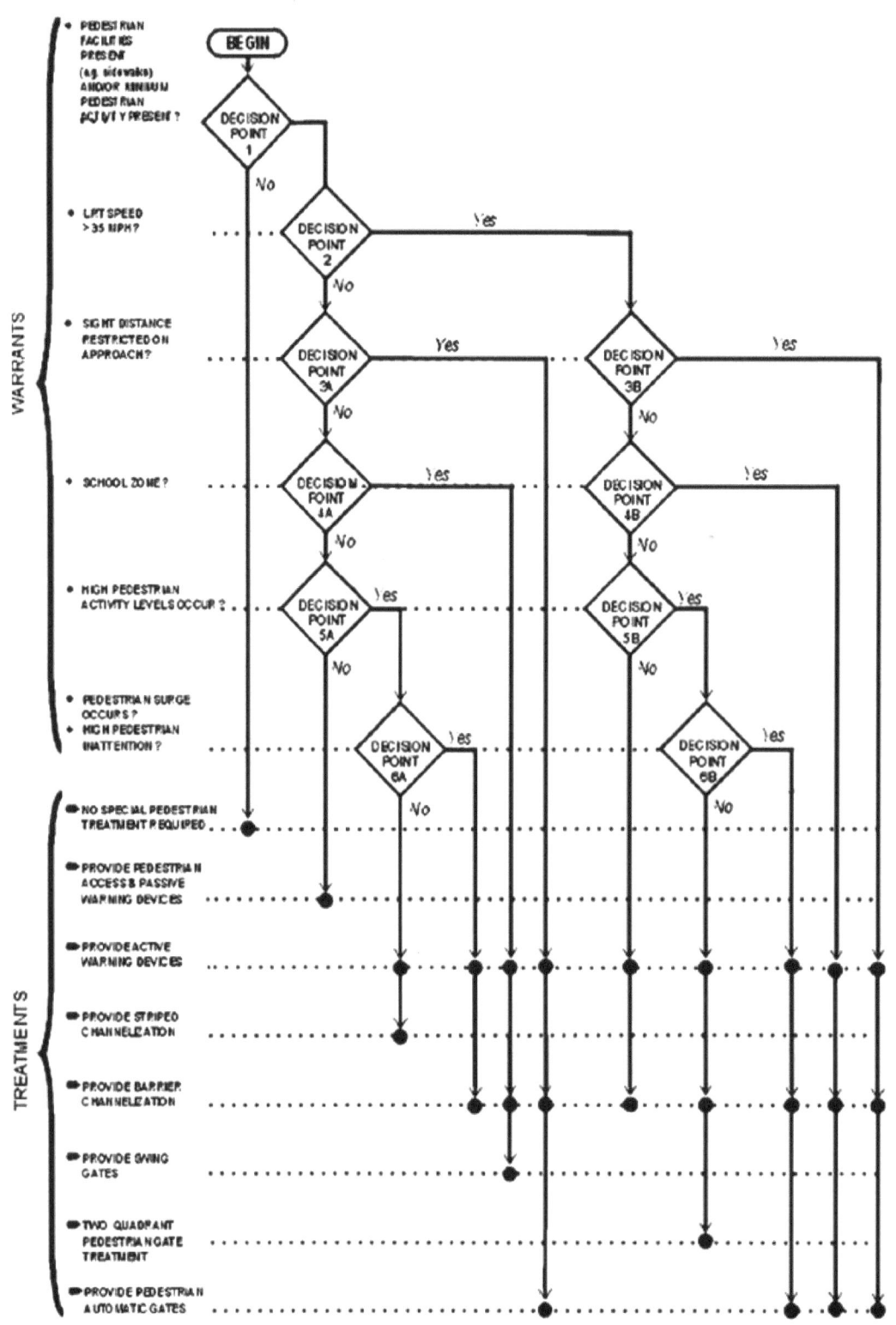

Figure 37. Pedestrian Controls Decision Tree [8]

In developing its Pedestrian Treatment Guidance Manual, the Utah Department of Transportation (UDOT) sets as goals to "establish a consistent procedure to evaluate" pedestrian crossings and to "establish a consistent application" for devices, yet acknowledges that no single standard

44

system is universal. However, it is likely that pedestrian safety at rail grade-crossings will benefit by increasing consistency in standards for treatments designed for this task [29].

4.3 Efficacy of Pedestrian Treatments

In 2013, Metaxatos found that the effectiveness of any particular sign or pedestrian treatment in "reducing the risk of a collision between a pedestrian and a train" is not known. The research also found that there are few existing methodologies allowing for assessing trade-offs among the various criteria used to select warning devices for deployment at pedestrian-rail grade crossings [22].

There has been some research performed for individual pedestrian treatments. Some of the research included:

- The effects of gate skirts on pedestrian behavior.

  A study performed at NJT, where pedestrian gate rushing was reduced by 78 percent while the gate was descending, only to have gate rushing increase by 12 percent while the gate was ascending [20] .

- A second train coming sign demonstration project at LACMTA.

  Risky behavior, in this case defined as pedestrians crossing tracks at less than 15 seconds in front of an approaching vehicle, was reduced 14 percent overall. Pedestrians crossing the tracks at 4 seconds or less ahead of the vehicle was reduced by 73 percent [37].

- The effects of an active another train coming system on pedestrian behavior.

  A study performed in Garfield NJ that showed pedestrian risky behavior statistically was unchanged in the presence of an active another train coming system. The results may be attributable to small sample sizes and extreme weather during the data collection [19].

Although there is some evidence that certain individual safety devices improve pedestrian outcomes, virtually no research has been performed on the efficacy of certain combinations for addressing the needs at particular types of pedestrian rail grade crossings.

5. Other Considerations

In a policy statement regarding pedestrian and bicycle accommodation, the USDOT encouraged State DOTs to consider "walking and bicycling as equals with other transportation modes" [38]. The policy statement included other recommended actions, such as

- Fostering increased use by bicyclists and pedestrians
- Accommodating people of all ages and abilities
- Improving non-motorized facilities during maintenance projects, an
- Going beyond minimum design standards

In 2011, United States Access Board (USAB) issued a Notice of Proposed Rulemaking (NPRM) titled Proposed Guidelines for Pedestrian Facilities in the Public Right-of-Way [24]. The new guidelines for public rights-of-way address issues such as access for blind pedestrians and wheelchair access and constraints posed by space limitations, roadway design practices, slope, and terrain.

Although many agencies have a policy to comply with accessibility requirements when designing and constructing pedestrian grade crossings, many do not. In addition, given the wide variety of grade crossing configurations and engineering issues, it may not be possible to upgrade all existing crossings to be fully compliant with ADA requirements.

The guidelines, when issued and legally enforceable, will impact future design considerations for pedestrian railroad grade crossings.

6. Conclusion and Recommendations

Based upon the Government framework of upcoming recommended guidelines for accessibility compliance and a Department policy statement regarding non-motorized users, there will be increased demand on grade crossings by a wider variety of users with a wider variety of abilities. Pedestrian safety techniques that work with one demographic may not readily transfer or be effective or accessible for another.

There are some studies that show efficacy of particular types of pedestrian treatments, such as the gate skirt study, but for the most part the effectiveness of any particular pedestrian grade-crossing treatment is not well documented.

Individual pedestrian treatments are seldom, if ever, used in isolation. They are used in combinations, as indicated by the decision trees presented in this document. It would be a practical approach, therefore, to study the effects of particular combinations of grade crossing treatments since that is the manner in which the treatments are deployed.

The recommendation of this study is to document particular types of pedestrian treatment combinations at various types of pedestrian crossings, and to quantify the efficacy through a before and after study of pedestrian violations and risky behavior. This approach differs from previous studies in that combinations of treatments will be analyzed rather than the individual treatments. The rationale for this is that pedestrian grade-crossing treatments are rarely, if ever, used in a stand-alone fashion, and that pedestrian behavior towards one treatment may be affected by other treatments at a particular grade-crossing.

Ideally, a study could be designed to document before and after pedestrian risky behavior activity at all, or as many as possible, grade-crossing improvement sites. The pedestrian risky behavior statistics, coupled with a complete before and after site inventory of treatments, would yield quantifiable results on what treatments are the most effective for keeping pedestrians safe.

New demands are being placed on pedestrian grade-crossings. New construction projects and upgrades will be required to conform to USDOT regulations regarding accessibility, once they are issued. Considerations for other users, such as bicyclists, must be incorporated as well.

7. References

1. Federal Railroad Administration, Office of Safety Analysis, http://safetydata.fra.dot.gov/OfficeofSafety/default.aspx viewed on 12/16/2014.

2. *2012 ROW Fatality Trespass Prevention Workshop,* http://www.fra.dot.gov/conference/trespass2012/ viewed on 9/18/2014

3. FHWA. (2009). *Manual on Uniform Traffic Control Devices.*

4. FHWA. (2007). *Railroad-Highway Grade Crossing Handbook.*

5. FRA. (2008). *Compilation of Pedestrian Safety Devices In Use at Grade Crossings.*

6. Mathieu, R. (2007). *Pedestrian Railroad Crossings Lessons Learned.* Presentation, Metrolink.

7. CPUC. (2008). *Pedestrian-Rail Crossings in California.* California Public Utilities Commission.

8. TCRP. (2001). *Report 69, Light Rail Service: Pedestrian and Vehicular Safety.*

9. *Caltrain Design Criteria.* (2011).

10. TRB. (n.d.). *Guidebook on Pedestrian Crossings of Public Transit Rail.* Retrieved Nov 7, 2014, from Transportation Research Board of the National Academies: http://apps.trb.org/cmsfeed/TRBNetProjectDisplay.asp?ProjectID=3324

11. Telephone interview with AMTRAK safety personnel, March 2014.

12. http://www.caltrain.com/Assets/_Engineering/engineering-standards/7000+Series+-+Grade+Crossings.pdf viewed 10/20/2014.

13. Telephone interview with Caltrain safety personnel, March 2014.

14. Telephone interview with Capital Metro safety personnel, March 2014.

15. PowerPoint Presentation entitled "NCTD Grade Crossing Pedestrian Treatments" received from NCTD on March 18, 2014.

16. Telephone interview with NCTD safety personnel, March 2014.

17. Telephone interview with Metrolink safety personnel, March 2014.

18. Telephone interview with NJT safety personnel, March 2014.

19. FRA. (2014). *Effect of an Active Another Train Coming Warning System on Pedestrian Behavior at a Highway-Rail Grade Crossing.*

20. FRA. (2013). *Effects of Gate Skirts on Pedestrian Behavior at a Highway-Rail Grade Crossing.*

21. Telephone interview with LIRR safety personnel, March 2014.

22. Metaxatos, P., & Sriraj, P. (2013). *Pedestrian/Bicyclist Warning Devices and Signs at Highway-Rail and Pathway-Rail Grade Crossings.* Chicago: University of Illinois at Chicago.

23. TCRP. (1996). Report 17, *Integration of Light Rail Transit into City Streets.*

24. Proposed Guidelines for Pedestrian Facilities in the Public Right-of-Way, https://www.access-board.gov/guidelines-and-standards/streets-sidewalks/public-rights-of-way/proposed-rights-of-way-guidelines viewed on 10/29/2014.

25. BikePortland http://bikeportland.org/2011/09/01/a-few-ideas-on-how-to-improve-streetcar-track-safety-58408, viewed 12/30/2014.

26. FRA. (2014). *Effect of Dynamic Envelope Pavement Markings on Vehicle Driver Behavior at a Highway-Rail Grade Crossing.*

27. The Source TransportatioN News and Views http://thesource.metro.net/2010/03/23/a-pedestrians-view-of-the-blue-line/ viewed on 1/07/2015.

28. SCRRA (2009). *SCRRA Highway-Rail Grade Crossings Recommended Design Practices and Standards Manual.*

29. Orange County Transportation Authority, http://www.octa.net/Metrolink/San-Clemente-Pedestrian-Crossings/ viewed 12/23/2014.

30. UDOT. (2013). *UDOT Pedestrian Grade Crossing Manual.*

31. Rosehill Rail, http://www.rosehillrail.com/products/anti-trespass-panels/specifications/ viewed 12/17/2014.

32. Siques, J. (2002). *Effects of Pedestrian Treatments on Risky Pedestrian Behavior.* Transportation Research Record 1793.

33. Metaxatos, et al. (2013). *Pedestrian and Bicyclist Safety at Railroad Grade Crossings.*

34. Information obtained from Volpe Center analysis of grade crossing activity at a grade crossing in Little Neck, NY (2013).

35. Capital Metro Railroad (2013). *Rail Systems Highway-Rail Grade Crossing Recommended Design Practices and Standards Manual.*

36. Fluor HDR. (n.d.). *At-Grade Crossing Design Guidance for Pedestrian Treatment Application.*

37. TCRP. (2002). Research Results Digest, No. 51.

38. FHWA. (2010). Policy Statement on Bicycle and Pedestrian Accommodation, http://www.fhwa.dot.gov/environment/bicycle_pedestrian/overview/policy_accom.cfm viewed on 12/17/2014.

Appendix A. Request for Information

U.S. Department
of Transportation

**Research and
Innovative Technology
Administration**

Volpe National Transportation Systems Center

55 Broadway
Cambridge, MA 02142

Engineering Strategies for Pedestrian Safety at Railroad Grade Crossings

Request for Information

The Volpe National Transportation Systems Center (Volpe Center), on behalf of the Federal Railroad Administration (FRA) is seeking information regarding engineering and design strategies for pedestrians at grade crossings. Engineering and design plays a major role in the prevention of pedestrian accidents at rail grade crossings. Many crossings use unique engineering and design strategies that may be applicable to other crossings for pedestrian safety. This information will help the Volpe Center with a current research study to identify areas where pedestrian compliance at grade crossings is high and engineering and design that may have contributed to the success. Volpe is particularly looking for the following information:

- Types of pedestrian safety engineering and design strategies
- Any data on their effectiveness in preventing trespasser activity and improving pedestrian safety

If you would like to volunteer any of this information to Volpe Center or if you have any questions, please contact Bernard J. Kennedy IV Bernard.Kennedy@dot.gov, 617.494.3591. Please feel free to transmit this request to others in the railroad and transit communities. The Volpe Center greatly appreciates your consideration of this information request.

Appendix B. Resources

The following reports contain useful guidance and documentation of pedestrian treatments for grade crossing applications:

Railroad-Highway Grade Crossing Handbook, Revised Second Edition, August 2007. U.S. Department of Transportation, Federal Highway Administration.

Manual on Uniform Traffic Control Devices for Streets and Highways, 2009 Edition. U. S. Department of Transportation, Federal Highway Administration.

Compilation of Pedestrian Safety Devices in Use at Grade Crossings, 2008. U.S. Department of Transportation, Federal Railroad Administration.

Transit Cooperative Research Program Report 17: Integration of Light Rail Transit into City Streets, 1996. National Research Council, Transportation Research Board.

Transit Cooperative Research Program Report 69: Light Rail Service: Pedestrian and Vehicular Safety, 2001. National Research Council, Transportation Research Board.

Transit Cooperative Research Program Report 137: Improving Pedestrian and Motorist Safety Along Light Rail Alignments, 2009. National Research Council, Transportation Research Board.

Rail Systems Highway Rail Grade Crossing Recommended Design Practices and Standards Criteria Manual, 2013. Capital Metro Railroad

SCRRA Highway-Rail Grade Crossings Recommended Design Practices and Standards Manual, 2009. Southern California Regional Rail Authority (Metrolink)

Pedestrian-Rail Crossings in California, 2008. California Public Utilities Commission

Appendix C. Literature Review Analysis

Title: **Pedestrian Railroad Crossings Lessons Learned**	Category: Pedestrian __X__ Grade Crossing _X_

Presentation briefly touches on initial pedestrian grade-crossing treatments as "sidewalk enhancements."
This presentation describes a brief history of pedestrian treatments for at-grade crossings. First, two alternatives to crossings are discussed: Above grade crossings, costing between $2 - $8 million (2007 USD) Undercrossings, costing between $1.5 - $3.5 million (2007 USD) The presentation goes on to describe the San Clemente Pedestrian trail, which successfully mitigated years of trespassing to control pedestrian traffic. Finally, pedestrian treatments at Pasadena Gold Line, Caltrain, Santa Clara Valley, LIRR, Metro-North Commuter Rail, and New Jersey Transit are discussed.
Study Length: N/A
Cost: N/A
Source: *Pedestrian Railroad Crossings Lessons Learned (2007)*, Presentation

Title: **Railroad-Highway Grade Crossing Handbook**	Category: Pedestrian _____ Grade Crossing _X_

Reference document on prevalent and best practices as well as adopted standards relative to highway-rail grade crossings.
The Special Issues section describes a number of treatments applicable to pedestrian crossings. Although the recommendations were for LRV, they appear to be applicable to highway-rail crossings in general. The recommendations include: Dynamic envelope markings – contrasting pavement texture to identify a vehicle's dynamic envelope through a pedestrian crossing, Curbside pedestrian barriers – bedstead barriers, fences, and/or bollards and chains to warn pedestrians where they should not go, Pedestrian automatic gates – as with vehicular automatic gates, installation in all 4 quadrants is recommended, Swing gates – gates that force pedestrians to pause and pull them open in order to enter the track area, Bedstead barriers – barricades placed in an offset manner that requires pedestrians to navigate a passageway through the barriers, Z-crossing channelization – a pathway whose design and installation turn pedestrians toward the approaching vehicle.
Study Length: N/A
Cost: N/A
Source: *Railroad-Highway Grade Crossing Handbook (2007), Federal Highway Administration*

Title: **Compilation of Pedestrian Safety Devices in Use at Grade Crossings**	Category: Pedestrian _____ Grade Crossing __X___
Report is a collection of both non- and MUTCD compliant devices in use by agencies and organizations.	
This document presents a variety of pedestrian safety treatments, including active warning devices, warning signs, gates, and channelization devices. While acknowledging that the MUTCD has the status of law with regards to traffic control devices, the report encourages the transportation community to participate in the MUTCD "Interpretations, Experimentations, Changes, and Interim Approvals" process so that agencies may continue to develop innovative pedestrian treatments.	
Study Length: N/A	
Cost: N/A	
Source: *Compilation of Pedestrian Safety Devices in Use at Grade Crossings, (2008) FRA.*	

Title: **Pedestrian-Rail Crossings in California**	Category: Pedestrian _____ Grade Crossing __X___
Recognizing an expansion of light rail transit and commuter rail systems, and the accompanying new stations and pedestrian-rail crossings, the California Public Utilities commission has created this guide for pedestrian-rail at-grade crossing design and improvements based on current industry practices.	

Document contains the following design elements
     Swing Gates
     Detectable Warnings
     Pedestrian Gates
     Flashing Light Assemblies
     Signage
     Crossing Surfacing
     Channelization
     Other Treatments
These treatments are also illustrated in the Design Examples section of the document.	

Study Length: N/A	
Cost: N/A	
Source: *Pedestrian-Rail Crossings in California, (2008) CPUC.*	

Title: **Manual on Uniform Traffic Control Devices**	Category: Pedestrian _____ Grade Crossing__X_
National standards governing all traffic control devices.	
The MUTCD sets minimum standards to ensure the uniformity of traffic control devices. The use of uniform TCDs (messages, location, size, shapes, and colors) helps reduce crashes and congestion. Uniformity also helps reduce the cost of TCDs through standardization. The information contained in the MUTCD is the result of either years of practical experience, research, and/or the MUTCD experimentation process. This effort ensures that TCDs are visible, recognizable, understandable, and necessary. The MUTCD is a dynamic document that changes with time to address contemporary safety and operational issues.	
Study Length: N/A	
Cost: N/A	
Source: *Manual on Uniform Traffic Control Devices, (2009) FHWA.*	

Title: **SCRRA Highway-Rail Grade Crossings, Recommended Design Practices and Standards Manual**	Category: Pedestrian _____ Grade Crossing__X_

Reference document on prevalent and best practices as well as adopted standards relative to highway-rail grade crossings.

The Special Issues section describes a number of treatments applicable to pedestrian crossings. Although the recommendations were for LRV, they appear to be applicable to highway-rail crossings in general.
The recommendations include:
Dynamic envelope markings – contrasting pavement texture to identify a vehicle's dynamic envelope through a pedestrian crossing,
Curbside pedestrian barriers – bedstead barriers, fences, and/or bollards and chains to warn pedestrians where they should not go,
Pedestrian automatic gates – as with vehicular automatic gates, installation in all 4 quadrants is recommended,
Swing gates – gates that force pedestrians to pause and pull them open in order to enter the track area,
Bedstead barriers – barricades placed in an offset manner that requires pedestrians to navigate a passageway through the barriers,
Z-crossing channelization – a pathway whose design and installation turn pedestrians' pathways towards a rail vehicle's approach.

Study Length: N/A
Cost: N/A
Source: *SCRRA Highway-Rail Grade Crossings, Recommended Design Practices and Standards Manual (June 2009),* Southern California Regional Rail Authority (Metrolink)

Title: **Caltrain Design Criteria**	Category: Pedestrian __X__ Grade Crossing__X_

Five pedestrian treatments were evaluated across three grade crossings along the Tri-county Metropolitan Transportation District of Oregon MAX light rail system in Portland, OR.

Regarding exiting crossings, Caltrain promotes the following approaches: closure, consolidation, enhancement of crossing, grade separation, and adaptation of new technology at crossings. Pedestrian treatments include pedestrian gate arms, pavement texturing, pavement marking, channelization (guard railing, fencing, swing gates)

Study Length: N/A
Cost: N/A
Source: *Caltrain Design Criteria(2011),* Caltrain.

Title: **TCRP Report 69: Light Rail Service: Pedestrian and Vehicular Safety**	Category: Pedestrian __X__ Grade Crossing__X_
This document presents the results of a study of 11 North American LRT systems where vehicles operate at speeds greater than 35 mph through crossings to improve safety.	
Presents the Pedestrian Controls Decision Tree flowchart and recommends its' use. A survey of 11 North American LRT systems found the following types of pedestrian warning and control devices in use: • Traditional Railroad Devices (bells, pedestrian automatic gates, flashing light signals) • Traditional Traffic Devices (pedestrian signal heads) • Customized Active Warning Devices (illuminated signs, with or without audio devices) • Modified Devices (pedestrian automatic gates with gate skirts)	
Study Length: N/A	
Cost: N/A	
Source: *TCRP 69: Light Rail Service: Pedestrian and Vehicular Safety(2001),* TRB.	

Title: **Rail Systems Highway-Rail Grade Crossing Recommended Design Practices and Standards Criteria Manual**	Category: Pedestrian __X__ Grade Crossing__X_
Document provides the highway grade crossing design criteria for the Capital Metro Rail signal system.	
Uses a modified version of the TCRP Report 69 decision tree, the Pedestrian-Rail Grade Crossing Design Consideration Flowchart. Identifies the following treatments for use at pedestrian-rail grade crossings: • Warning devices • Channelization • Passive Devices (signage, pavement markings, swing gates) • Active Devices (pedestrian gates)	
Study Length: N/A	
Cost: N/A	
Source: *Rail Systems Highway-Rail Grade Crossing Recommended Design Practices and Standards Criteria Manual (2011),* Capital Metro Railroad.	

Title: **Utah Statewide Pedestrian Treatment Guidance Manual Standards for At-Grade Crossings**	Category: Pedestrian __X__ Grade Crossing__X_
A presentation given at the ITE 2013 Annual Meeting.	
UDOT presents their goal of developing a statewide pedestrian treatment guidance manual. The manual will establish a consistent procedure to evaluate pedestrian crossings, establish a consistent application for control devices, and develop standard drawings. Presents a Pedestrian Grade Crossing Flowchart to assist in determining pedestrian treatments. The flowchart consists of two versions: one for urban pedestrian crossings and one for rural.	
Study Length: N/A	
Cost: N/A	
Source: *Utah Statewide Pedestrian Treatment Guidance Manual Standards for At-Grade Crossings (2013),* UDOT.	

Title: **At-Grade Crossing Design Guidance for Pedestrian Treatment Application**	Category: Pedestrian __X__ Grade Crossing__X_
Documents the methodology for at-grade pedestrian treatments at highway-rail grade crossings.	
Using the decision tree presented in TCRO Report 69 and a version modified by CPUC, a project-specific decision tree was developed. Some of the customizations include and increased distance to schools, a decision point for multiple tracks, and a removal of Swing Gates as a potential treatment.	
Study Length: N/A	
Cost: N/A	
Source: *At-Grade Crossing Design Guidance for Pedestrian Treatment Application (nd),* Fluor HDR.	

Appendix D. Data

Grade Crossing Incidents by Type of Vehicle

| | A | B | C | D | E | F | G | H | J | K | M | |
Year	Auto	Truck	Truck-trailer	Pick-up Truck	Van	Bus	School Bus	Motorcycle	Other Motor Vehicle	Pedestrian	Other	Total
1994	2952	1241	544	0	0	3	3	22	0	77	157	4999
1995	2714	1192	506	0	0	3	3	14	0	74	143	4649
1996	2469	1102	471	0	0	8	4	11	0	95	108	4268
1997	2080	681	490	335	96	10	1	7	49	73	43	3865
1998	1813	460	477	449	115	3	4	7	56	91	46	3521
1999	1773	408	479	515	131	6	1	7	47	86	59	3512
2000	1753	415	461	559	164	4	4	12	67	88	62	3589
2001	1516	350	465	523	129	7	3	6	65	92	81	3237
2002	1449	338	454	492	141	4	3	8	45	71	76	3081
2003	1428	274	393	488	139	8	0	12	131	85	19	2977
2004	1433	302	441	498	116	4	4	11	143	111	22	3085
2005	1411	237	509	463	144	3	1	15	129	115	39	3066
2006	1298	225	508	494	120	6	2	7	146	102	34	2942
2007	1160	203	492	472	131	2	1	8	161	110	38	2778
2008	1046	132	376	423	105	4	1	10	166	130	36	2429
2009	879	109	269	306	72	5	1	5	138	112	37	1933
2010	925	140	304	308	73	4	0	4	111	144	39	2052
2011	955	144	352	269	53	2	0	5	93	116	49	2038
2012	866	155	333	265	64	3	1	6	97	99	49	1938
2013	961	170	336	249	58	2	0	6	96	117	48	2043
2014	765	128	270	210	47	0	0	5	77	97	47	1646
Total	31646	8406	8930	7318	1898	91	37	188	1817	2085	1232	63648

Grade Crossing Total Fatalities by Type of Vehicle

| | A | B | C | D | E | F | G | H | J | K | M | |
Year	Auto	Truck	Truck-trailer	Pick-up Truck	Van	Bus	School Bus	Motorcycle	Other Motor Vehicle	Pedestrian	Other	Total
1994	382	138	15	0	0	0	0	9	0	50	23	617
1995	328	143	22	0	0	4	7	6	0	47	24	581
1996	269	121	21	0	0	0	0	3	0	60	13	487
1997	246	89	21	28	18	0	1	2	13	38	4	460
1998	205	57	13	61	21	0	2	2	4	53	12	430
1999	185	33	23	72	22	0	0	5	5	50	12	407
2000	179	43	19	78	32	0	3	1	7	51	13	426
2001	174	40	20	76	26	0	0	1	8	67	9	421
2002	176	33	10	56	26	0	0	1	8	35	12	357
2003	144	26	11	61	16	0	0	5	18	50	3	334
2004	150	26	17	64	11	0	0	4	17	73	9	371
2005	138	23	21	68	19	0	0	0	15	58	17	359
2006	132	25	22	84	22	0	0	0	20	53	11	369
2007	113	27	18	54	19	0	0	2	32	59	15	339

2008	115	9	9	52	9	0	0	2	26	64	5	291
2009	108	17	5	29	10	0	0	1	12	60	7	249
2010	82	19	7	38	7	0	0	3	13	82	12	263
2011	90	15	13	35	8	0	0	2	8	68	11	250
2012	87	8	15	28	5	0	0	1	21	56	12	233
2013	86	14	12	22	10	2	0	2	14	65	16	243
2014	56	13	9	27	3	0	0	0	6	57	15	186
Total	3445	919	323	933	284	6	13	52	247	1196	255	7673

Accident Data downloaded on December 15, 2014

Accident data current through September 30, 2014

Pickup truck, van and other motor vehicle fields were added to the database in 1997

Total fatalities include highway users, railroad employees and passengers

Suicide and attempted suicide data were collected from June 1, 2011

Abbreviations and Acronyms

ADA	Americans with Disabilities Act
CMTA	Capital Metropolitan Transportation Authority
CPUC	California Public Utilities Commission
FHWA	Federal Highway Administration
FRA	Federal Railroad Administration
LACMTA	Los Angeles County Metropolitan Transportation Authority
LED	Light Emitting Diode
LIRR	Long Island Railroad
LRT	Light Rail Transit
LRV	Light Rail Vehicle
MPH	Miles Per Hour
MUTCD	Manual of Uniform Traffic Control Devices
NCTD	North County Transit District
NJT	New Jersey Transit
NPRM	Notice of Proposed Rulemaking
OL	Operation Lifesaver
RFI	Request for Information
ROW	Right-of-way
SCRRA	Southern California Regional Rail Authority (aka Metrolink)
SWS	Smart Warning Systems
TAMS	Train Approaching Message System
TCRP	Transit Cooperative Research Program
TRB	Transportation Research Board
TTI	Texas A&M Transportation Institute
UDOT	Utah Department of Transportation
USAB	United States Access Board
USDOT	United States Department of Transportation

www.ingramcontent.com/pod-product-compliance
Lightning Source LLC
Chambersburg PA
CBHW051050180526
45172CB00002B/585